SINK OR FLOAT

A Sesame Street Science Book

Marie-Therese Miller

Lerner Publications ◆ Minneapolis

Sesame Street has always been a community of curiosity and exploration. We know that all children are naturally curious about the world around them. Understanding basic science concepts can be fun for everyone in the neighborhood—including kids! In the Sesame Street® World of Science books, *Sesame Street*'s favorite furry friends help young readers learn about how the world works.

Sincerely,

The Editors at Sesame Workshop

Table of Contents

SINK AND FLOAT

Some things sink in water. Others float. To sink is to fall below the top of the water. To float is to stay on or near the top of the water.

HOW OBJECTS SINK OR FLOAT

Objects are made of tiny particles. You need a microscope to see particles.

Me can't see particles that make up cookies. But cookies still yummy!

Some objects have particles that are tightly packed. These have a higher density than objects with loosely packed particles. Density is how close together particles are.

Things sink when they have a higher density than water.

Things float when they have a lower density than water.

An object has buoyancy if it can float on water.

My basketball floats on water.

Let's predict whether some things will float or sink. When you predict, you say what you think will happen.

Someone drops ice cubes in a glass of water. Do you predict the ice cubes will sink or float?

I like to drink ice water!

Ice cubes float. Water particles spread out when they freeze into ice. Ice is less dense than water.

Someone drops a coin in a fountain and makes a wish. Coins are small and dense.

Will the coin sink or float?

Coins sink in water. They are denser than water.

A person puts a sailboat in the water. Will it float or sink?

The answer rhymes with boat!

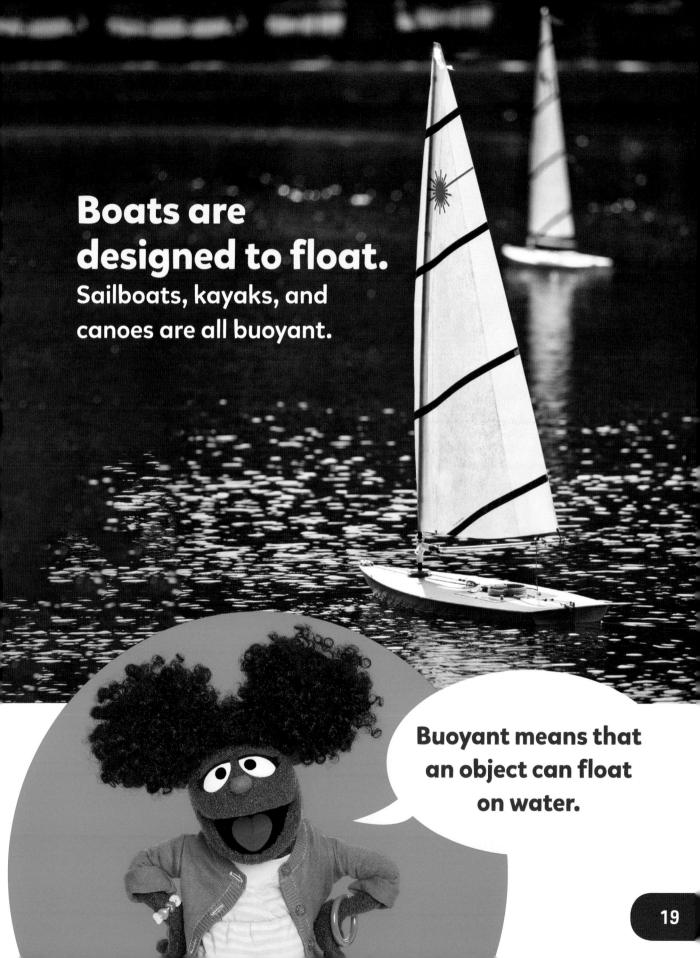

Boats are designed to float.
Sailboats, kayaks, and canoes are all buoyant.

Buoyant means that an object can float on water.

Here is a tricky question:

Do balls float or sink in water?

There are many different kinds of balls. I love playing with soccer balls!

Some balls float. Others sink.

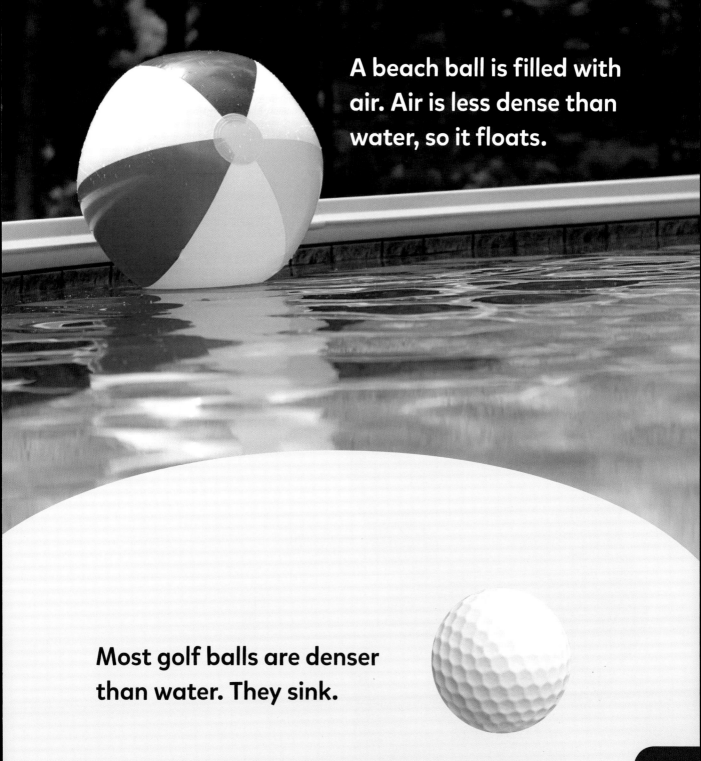

A beach ball is filled with air. Air is less dense than water, so it floats.

Most golf balls are denser than water. They sink.

Will marshmallows sink or float in water? Make a prediction.

I love marshmallows in my hot cocoa.

Marshmallows float on water. They float in hot cocoa too! Marshmallows have air in them. This makes them less dense than water—and hot cocoa.

Do you predict feathers will float or sink in water?

My feathers are very light . . . and very yellow!

A feather floats! It is less dense than water.

All kinds of objects sink or float. What can you find that sinks or floats? Have a grown-up help you!

Science All Around

Make a prediction about what will sink and float. Check your answers on page 30.

Will a carrot **sink** or **float** in water?

Ask a grown-up to help you!

Do you predict an apple will **sink** or **float** in water?

Elmo wonders if green apples will float!

Glossary

buoyancy: able to float on water

density: how closely packed particles are in an object

float: to stay on or near the top of the water

particles: tiny parts that make up objects

sink: to fall below the top of the water

Answers to Science All Around

Carrots sink because they are denser than water. Apples float because they are less dense than water.

Learn More

Miller, Marie-Therese. *States of Matter: A Sesame Street Science Book*. Minneapolis: Lerner Publications, 2023.

Wilder, Nellie. *Staying Afloat*. Huntington Beach, CA: Teacher Created Materials, 2019.

Zalewski, Aubrey. *Sink vs. Float*. Mankato, MN: Child's World, 2020.

Index

Photo Acknowledgments

Image credits: Kokhanchikov/Moment/Getty Images, p. 4; RichLegg/E+/Getty Images, p. 7; SDI Productions/E+/Getty Images, p. 8; Nikki Zalewski/Shutterstock, p. 10 (top); scanrail/iStock/Getty Images, p. 10 (bottom); Chones/Shutterstock, pp. 11, 21 (bottom); Jonathan Kirn/The Image Bank/Getty Images, p. 12; Backyard Productions/Alamy Stock Photo, p. 15; LumineImages/Shutterstock, p. 17; Bart Conrad/iStock/Getty Images, p. 19; Margaret M Stewart/Shutterstock, p. 21 (top); Cavan Images/Getty Images, p. 23; Anton Starikov/Shutterstock, p. 25; DGLimages/iStock/Getty Images, p. 26.

Cover: DGLimages/iStock/Getty Images.

With appreciation to all scientists and engineers, particularly John, Michelle, Meghan, John Vincent, Erin, Elizabeth, and future scientist, Greyson

Lerner Publications Company
An imprint of Lerner Publishing Group, Inc.
241 First Avenue North
Minneapolis, MN 55401 USA

For reading levels and more information, look up this title at www.lernerbooks.com.

Main body text set in Mikado.
Typeface provided by HVD.

Editor: Amber Ross **Designer:** Mary Ross
Lerner team: Martha Kranes

Library of Congress Cataloging-in-Publication Data

Names: Miller, Marie-Therese, author.
Title: Sink or float : a Sesame Street® Science Book / Marie-Therese Miller.
Description: Minneapolis : Lerner Publications, [2023] | Series: Sesame Street ® World of Science | Includes bibliographical references and index. | Audience: Ages 4–8 | Audience: Grades K–1 | Summary: "Why do some things sink and others float? Simple text introduces readers to key science curriculum, and commentary from the Sesame Street friends make the learning fun"— Provided by publisher.
Identifiers: LCCN 2022006366 (print) | LCCN 2022006367 (ebook) | ISBN 9781728475806 (lib. bdg.) | ISBN 9781728486178 (pbk.) | ISBN 9781728484822 (eb pdf)
Subjects: LCSH: Floating bodies—Juvenile literature. | Matter—Properties—Juvenile literature.
Classification: LCC QC147.5 .M55 2023 (print) | LCC QC147.5 (ebook) | DDC 532/.25—dc23/eng20220708

LC record available at https://lccn.loc.gov/2022006366
LC ebook record available at https://lccn.loc.gov/2022006367

Manufactured in the United States of America
1-52147-50610-6/30/2022